# REMOVING SKIN TAGS, WARTS AND MOLES SAFELY AND NATURALLY

Written by
**Jimmy J. Jacks**

"To Do Is To Achieve
To Achieve Is To Reap Reward"
-Dennis Killian

# Table of Contents

# Introduction

What do I have here, you ask.  You know it's some type of growth.  It's a mole . . . no wait . . . it's a wart . . . oh, no, it's not.  I recognize it now -- it's a skin tag!  Okay, so maybe I have no idea what you have there.  But I do know it's a skin growth of some type.

Does this sound like any conversation you've had recently with friends and family?  You've got a growth and you're not sure how to classify it.  But more than classifying the darned thing, you'd just like to be rid of it once and for all.

Oh yeah, the doctor tells you it's one thing and sends you on your way with a prescription in hand.  Meanwhile, you're still not quite sure what type of growth it is.  And your confidence that this little slip of paper can transform itself into an effective remedy is waning the further you get from the doctor's office and the closer you get to the pharmacy.

**Bingo!**  Your fears were right.  The prescription didn't work.  Now, here you are several months later, reminding your doctor of what he did and asking for something that might work.

Before you undergo any more conventional therapy, why not read this book?  It's chock full of information on moles . . . warts . . . and skin tags.  And it was written out the same frustration that brings you to it now.

Yes, I too had several warts as well as some moles hanging around.  Then my doctor told me I even had some skin tags. How lucky could a person be, I mused. I'm blessed with all three of the growths.

I was having difficulty telling one from the other, and I certainly couldn't explain in a simple way what any of the three really were.

So I went routing around the internet to see what all of the major sites had to say. I got quite an education, but it took forever. I decided that I would compile all the information I found in one easy to use e-Book. Not only would I include a variety of home, herbal and nutritional remedies, but I would explain in everyday terms what these growths were.

I wanted anybody who read this book to understand with ease the differences between a wart, a mole and a skin tag. I wanted readers to be able to turn to a specific area immediately, if they wanted to, and start scouring the page for the removal process that best suited them.

And, so, out of that intent this book was born. You'll find the organization of this book quite simple. The first chapter describes the three different growths. Chapter 2 gives you an idea of the options your personal health care physician has at their disposal to treat and remove these three growth.

And then we get to the meat of the book and what you've no doubt been waiting for -- natural ways to remove warts, moles and skin tags safely and naturally in the privacy of your own home.

One of the first things you notice is that the methods by which moles and warts are treated by both conventional doctors and in the natural remedies chapters are very similar. And that they are -- at least in the case of how to remove them. Just about every remedy you see for warts also works for moles, unless I've stated differently. Skin tags, while similar, are just different enough to deserve their own section (how special is that?).

I've not only included herbal solutions, but remedies that involve vitamins and minerals as well as removal choices using items you probably have lying around your home -- like duct tape. (No, I'm serious. Wait until you read it!)

The last chapter, while not directly related to removal, deals with your being able to recognize when a good mole has gone bad. As you're probably aware, an irregular growth on your skin may indicate the presence of skin cancer. This short chapter is just long enough to remind you of the symptoms and urge you to get early treatment.

I'm sure you're eager to begin your search for safe and easy removal of moles, warts and skin tags. I'd hate to keep you waiting any longer.

# Here's looking at you, kid!

# Chapter 1:
## Moles, Warts, Skin Tags:
## What Are They?

So you're looking at your hands wondering, just where in the world that wart came from. You think back on your activities from the last several days. No, you just can't seem to remember handling a toad, so how did you get that wart?

Let's dispel that myth once and for all. Even if you did handle a toad recently, that poor creature had nothing to do with you developing a wart.

Still, that leaves you wondering about your unsightly growth. Why are you singled out for this skin anomaly? Quite frankly, you're not being "singled out". The truth of the matter is that warts affect just about everyone.

Warts are second only to acne as being the most common dermatological complaint. In the U.S. alone, nearly $125 million is spent every year on the treatment of warts! That's a lot of warts!

According to Robert Garry, an associate professor of microbiology and immunology at Tulane University School of Medicine, approximately 10 percent of the population has at least one wart.

To break this down a little farther, he is quoted in Rodale's **The Doctor's Book of Home Remedies** as estimating that at any one time, 75 percent of the population will get at least one wart during their lifetime.

So it seems pretty inevitable that somewhere along the way, your turn to battle the wart will come. The more you know about what you're battling, the better your odds of success.

In addition, moles and skin tags are also common complaints and in some ways related. Conventional treatments for the three of these are very similar. Even home remedies that work to rid your skin of that wart can also get rid of the mole as well. There's even some overlap with self-treating for the skin tag.

## BUT EXACTLY WHAT IS A WART?

A wart appears as a skin growth caused by any one of between 60 and 80 strains of papillomaviruses. They are technically benign tumors of the epidermis -- or skin. The offending virus enters the body through a break in the skin itself.

While they can develop on a person regardless of their age, warts are most commonly found in children. Seldom are they found on older individuals.

If you pay close attention, you'll discover that this virus likes nothing better than warm, moist environments. Oh, you know the type. Those warm places such as locker rooms or even your shoes when your feet perspire and there's no where for that perspiration to go.

And yes, warts really are contagious. But more often than not, they're contagious on a single individual. This is because the virus that causes the growths only spreads through prolonged and repeated contact. They're seldom spread from one person to another.

Not only that, but the virus that causes warts only spreads through prolonged and repeated contact. Of course, the one exception to this is genital warts.

For the most part, warts are indeed bothersome but harmless. And believe it or not there really are different types of warts. There are **mosaic warts,** which grow together in clusters and **single warts** which show up as isolated growths.

Additionally, the medical community classifies warts according to their location and their shape. Common warts -- which carry the scientific name of **verrucae vulgaris** -- are just that. These are the common, "garden" variety kind that just about everyone gets occasionally.

These are best described as firm growths that more often than not are characterized by a rough surface. The shapes of these warts vary. Sometimes you'll discover they are nearly perfectly rounded. Other warts may have various irregular shapes.

Along the same vein, the colors of the warts may be different from person to person and even from wart to wart. While black and dark brown are probably what you recognize as common colors first, it's not uncommon for some of the growths to appear gray, or even yellow.

Warts, for all the misery they cause us, are actually physically quite small. They are usually not much larger than a half inch in diameter.

## WHERE YOU CAN FIND THEM

If you suffer from these growths then you already know the answer to this already: everywhere and anywhere. But many times warts pop up on areas that are frequently injured.

For example, many people get warts on their knees, fingers and even around the nails. In fact, warts that appear exclusively around the nails are referred to as **periungual warts.**

Common warts which can be found anywhere on the body have the potential to spread to the surrounding skin.

**Plantar warts**, as you probably already know, are those that appear on the soles of the feet. Here, they're very often flattened by the pressure involved in walking.

Many times you'll notice they're also enveloped by thick-ened skin. This variety is characteristically unique be-cause they are normally hard as well as flat. Their sur-face is rough and they possess well-defined boundaries.

These warts can even be found at times on the top of the foot. Here they usually appear fleshier and are usu-ally raised. Plantar warts are usually gray or brown with a small black center. Additionally, they have a unique characteristic. They may bleed from many very tiny spots -- similar to pinpoints -- when the surface is shaved or cut while receiving medical treatment.

## STILL MORE WARTS

Think we're done describing warts? Think again. Not quite. Ever hear of a wart called **filiform?** This specific type of wart is long and narrow. More often than not, they're found on the eyelids, as well as on the face, neck or even the lips.

Then there are **flat warts**, which most commonly ap-pear on children and young adults. These usually mani-fest in groups. Another telltale characteristic of this type of wart is the color. Many times these are pink or even flesh-colored. Sometimes they may be yellow-brown.

For the most part, flat warts appear on the face as well as on the top of the hand. Other locations this type of wart likes to appear is in the beard area of men and on the legs of women. By the way, it's not unusual for the act of shaving these areas to spread them to other parts of the body.

Finally, **genital warts** -- or **venereal warts** -- can be found on the penis, anus, vulva, vagina and the cervix. These warts are generally shaped irregularly. They're bumpy growths that are very often described as having the texture of . . . well . . . cauliflower! I'm afraid it's true. More than one person has described them like this!

Except for plantar warts, all of these growths are painless. Plantar warts, depending on their location, can become very painful when you place pressure on them – such as when you're walking.

---

# Moles:
# Test Your Knowledge

*Here's a quick quiz: We are all born with all the moles we will ever develop. True or false?*

*If you said true you'd be right. At birth, many of them are not visible. But as the child ages, these moles darken and become more apparent.*

---

## THEN WHAT'S A MOLE?

One of the most common skin abnormalities, a skin mole is a round or oval spot on the skin with a coloring that can range from pink to red to brown or even black. With the possibility of moles appearing on any part of the body, you can statistically depend on finding between 10 and 50 moles on your body.

Medically speaking, a mole is called a nevus. And if your physician speaks of "nevi", then he's talking about more than one mole.

A mole is actually nothing more than a collection of cells called melanocytes. These cells are naturally present in all areas of the skin as a part of its pigmentation. It's when these cells form in large enough numbers that eventually a mole appears.

If you examine your moles closely enough, you'll notice that you not only have some that are flat, but a few that are raised as well. In fact, if you take a real close look, you'll discover quite a diversity of moles. A few may have a hair coming out from them. This is quite normal, by the way.

If you have a mole or two that just bother you, consider getting them removed. Moles are so common that many individuals don't even think about removing them. But sometimes they develop in strategic areas which you may not feel comfortable with. That's when you may consider getting rid of them.

## SKIN TAGS?

If you have never seen one, you probably don't realize what they look like. The name, though, is a pretty good description of their appearance. A skin tag is an ac-quired, benign growth of skin that looks like a small piece of soft, hanging skin.

Skin tags are best described as bits of skin-colored or flesh-colored tissue projecting from the surrounding skin from a small, narrow stalk. You may have heard these growths referred to as skin tabs or even barnacles.

Medically speaking, they're known by several different names as well. You may hear your dermatologist refer to these acrochordons, Pappilomas, soft fibromas or even pedunculated filiform. This last term indicates that the growth is really on a stalk and is threadlike.

Actually quite harmless in nature, these tags seem to accumulate more on some individuals than others. Some people may have as many as 50 to 100 tags on their body.

What causes this growth isn't really clear. It appears that it may have something to do with the friction creat-ed by the act of rubbing parts of the body together or tight clothes rubbing certain areas of the body. That's why sometimes gaining weight may cause the growth of tags. Then again, some individuals have an abundance of these solely due to genetics.

Women are just as likely to get these as men. In some individuals even a moderate increase in weight can trigger a dramatic increase in these growths.

Skin tags also appear on those individuals of normal weight but who have larger breasts. The tags themselves usually form under the breasts. When formed here, many times the cause is due to the presence of an underwire bra that has been continually rubbing in this area.

They also tend to form in the underarm area, at the base of the neck on the eyelids and even in the folds of the groin. In fact, in teens the underarm area is a common area for the tags. It's thought they're created due to rubbing from the various sports teens are usually involved in.

If none of these conditions fit your profile, yet you've developed skin tags, think back to when they first appeared. It's not unusual for these growths to be stimulated by hormonal changes in a body. Specifically, women who are pregnant develop them.

The beginnings of a skin tag may be quite small, even as small as a flattened pinhead-sized bump. While most tags are not larger than 5 millimeter in diameter, some grow as large as the size of a big grape -- about a one centimeter in diameter or even a fig, about five centimeters.

## SYMPTOMLESS

For the most part skin tags produce no symptoms. They usually aren't painful. The possible exception to this is if they are repeatedly irritated. This could be caused by the rubbing of clothing or by the chafing of surrounding skin

Not only can this hurt, it may actually cause bleeding, especially if the tag gets twisted on its stalk. In this case, a blood clot may form within the small tag itself, which makes the entire area very painful. Not only that, but the presence of the clot turns the color of the tag into a purple or a black.

These tags may very well fall off on their own after about three (3) to ten (10) days. Technically called **thrombosed skin tags**, they don't require any additional treatment.

But don't worry when it comes to your long-term health since skin tags are neither cancerous nor precancerous growths.

## SKIN TAG DOPPELGANGER

Yes. It's possible that what you're calling a skin tag is really a different growth "masquerading" as a skin tag. Well, let's just say that you've mistakenly identified a growth as a skin tag when it really is something different.

And what could possibly be mistaken as skin tags? How about seborrheic keratosis. You might know these under the more generic name of friendly barnacles. Even common moles and warts may be initially mistaken for skin tags.

Now that we've identified moles, warts and skin tags, you have a better idea of what type of growth you're about to show your physician when you visit.

The next step is deciding what to do with these growths. The following chapter gives you some idea of how your doctor may treat them.

# Chapter 2:
## Removing Growths With Your Doctor's Help

Surprisingly, many warts -- and this is especially true for the common variety -- disappear on their own about a year or two after they debut. When they disappear naturally like this, there's seldom even a telltale mark that they were ever there. In these cases, treatment is not necessary.

Genital warts, by contrast, are far more likely to linger. But more than that, they're also more contagious. For this reason doctors usually remove them or treat them with drugs.

Of course, removal of the wart doesn't necessarily guarantee that the wart won't return at some point in the future. Many times, in fact, once a wart has been medically removed, it returns. Among the most difficult to get rid of is the plantar wart.

## HOW DO YOU GET RID OF A WART?

There are various methods and approaches of removing warts, not the least of which is to cut them. Sometimes a physician opts to freeze them or, going the other extreme, burns them with either laser or an electrical current.

Dermatologists call it **cryotherapy.** That's when they freeze the wart in order to remove it. We're not talking about sticking your fingers in the freezer section of your grocery store though. Here, freezing involves using substances call **cryogens.**

The most common of these substances is **liquid nitrogen.** But your dermatologist may use the older substance called **carbon dioxide snow.** This particular method has been around more than 20 years. For the most part it's been replaced by the liquid nitrogen, but some health care providers still use it.

Another cryogen is **dimethyl ether** and propane, known as DMEP. This is actually what the over-the-counter treatment, Wartner, contains.

In the long term, this may actually be the best and most effective treatment of skin lesions. It is inexpensive, relatively speaking. But beyond that, it's extremely safe and quite reliable.

The only real concern here is that the lesion is properly diagnosed. Cryotherapy should not be used on cancerous areas or any growths that have yet to be diagnosed but which could be melanoma.

## So what's involved?

If your dermatologist is using liquid nitrogen it means that they'll be freezing your wart down to a temperature of -196°C. Just in case you didn't read that properly, let me report that for you: "negative 196." That's downright chilly!

The nitrogen is applied to the skin with a cotton-tipped applicator or instruments that are either called a **cryospray** or a **cryoprobe.** As you can probably guess with temperatures that cold, the nitrogen only needs a few seconds of contact with the growth. The actual freeze period depends on the size of the lesion being treated as well as the depth of the roots of the lesion. Obviously, the deeper the dermatologist feels he needs to go to ensure full removal, the longer he'll keep the nitrogen on the spot.

In some cases, you may have to return for a second session. These are called **"double freeze-thaw,"** which is usually only performed on skin cancers or in the case of very stubborn, resistant viral warts.

If your doctor uses the carbon dioxide snow, then a different method is executed. The first difference is the temperature. Carbon dioxide snow only yields a temperature of approximately -78.5° Celcius. Sometimes he'll combine this with acetone to create a slush-like mixture which is applied directly to the growth.

Of all of the freezing techniques, though, DMEP is the warmest -- relatively speaking -- coming in at a mere -57°C. This substance is found in an aerosol can. In fact, you can buy it yourself without a prescription. Using a foam applicator, the DMEP is pushed onto the lesion for a maximum of 40 seconds, depending on the specific size of the wart or mole being treated.

## POTENTIAL HAZARDS OF CRYO-THERAPY

As you might imagine Cryotherapy may be painful, not only at the time of the treatment, but for a period afterwards as well. At the very least this treatment will sting you.

Immediately following the treatment, the area may swell and get red. Some of these symptoms may be mitigated simply by applying a topical steroid immediately following the freezing. You may also want to take aspirin to help reduce both the inflammation and the discomfort of the therapy.

## POST-TREATMENT CARE

Once the treatment has been given, that's not the end of the story. You'll notice a series of events starting shortly after the cryotherapy treatment is completed. The first is a blistering of the area.

In some cases, you'll develop a clear blister, but don't be surprised if it's red or even purple due to bleeding.

This is normal and is not an indication of something gone awry.

During this phase of the post-therapy, the blister needs no special treatment while it's healing.  The most you'll want to do with this is to gently wash it at least once and preferably twice a day.

Within a few days, the blister heals over, creating a scab.  The scab should be dabbed with petroleum jelly. Don't try to speed your body's healing process along by picking at the scab in an attempt to remove it sooner than necessary.  It'll come off in a matter of 10 days or so if it's located on the face.  If you've had warts or moles removed from your hands, it may take up to three weeks for the scab to fall off by itself.

Don't become alarmed if a scab located on the lower portion of your leg takes longer to heal.  For some individuals a scab may remain for as long as three months. This, actually, is not unusual.  It's just that healing in this area of the body takes longer.

## SECONDARY INFECTION?

The development of a secondary infection brought on by cryotherapy is practically unheard of.  In the rare instances it does occur, it causes increased pain, along with swelling.  A secondary infection can also be recognized because it produces a thick, yellow blister filled with pus. You may also notice redness around the treated area as a sign of the presence of a secondary infection.  If you do notice these symptoms, you should consult with your doctor.  It may be necessary to take an oral antibiotic or to use a topical antiseptic to combat it.

If your treatment involved an area close to the eye, then you may develop a puffy eyelid for a while.  This may not develop until the following morning.

# AND THE FINAL ANSWER IS . . .

A blemish-free area! For the most part, cryotherapy leaves no telltale marks behind it. If, however, your particular growth had to be treated for a prolonged length of time, or the freezing needed to occur deeper than most growths, you may notice a mark.

Some people end up with a white mark where the wart once was. This is called **hypopigmentation.** In some cases an actual scar may occur.

Even if you get only a "light freeze", a white mark may still form following the healing process. If you have a naturally dark complexion, then this may be noticeable. While the mark fades with time for the most part, for some individuals its presence is permanent.

Another side effect of cryotherapy may be numbness. This is especially true if your treatment involved the side of one of your fingers. This is only temporary. Your feeling in this area should return within a few weeks. Don't be alarmed if it takes several months, though, as some individuals have reported.

## LASER REMOVAL

It's no use, you say. You've tried everything. Still nothing removes that stubborn wart. Say hello to the **pulse dye laser.** This just may be the answer to your prayers. It's a relatively new approach to wart removal. And it specializes in removing the normally persistent growth. And while new, it comes bearing a relatively impressive track record.

The pulse dye laser has already been known to remove up to two thirds of the growths caused by the human papilomavirus. Odds are good, then, that yours can be included in one of them.

The procedure is simple enough. The laser energy is focused on the wart tissue itself. Surprisingly, the vast majority of the laser light actually bypasses the upper layers of the skin, known as the epidermis. Instead it soaks down and is concentrated onto the tiny blood vessels feeding the growth. This is near what's called the "dermis" of the skin.

During this procedure it's not unusual to feel a slight stinging, much like being hit with a spurt of bacon grease while you're making breakfast. For those who are sensitive to pain, your doctor may offer to use a topical or a local anesthesia on the area. For the most part, though, this is unnecessary.

Immediately following treatment, the area affected swells slightly and it may turn a blush color. This is normal and within several days this heals. The result is that the texture and tone of your skin returns to normal.

The good news about this treatment is the relative short period of time it takes. Depending on the number of warts or moles being removed in a single session, it may take from 10 to 30 minutes.

The actual number of laser sessions you'll need depends again on the number of warts and their individual "stubbornness". Simple warts usually require, at most, two sessions. Many of these in fact, can be removed with a single laser application. The deeper warts and those more resistant may take several more treatments.

If you're thinking that this treatment sounds as if it carries a great deal of risk, think again. It is the only treatment approved by the U.S. Food and Drug Administration for being used on children.

The risk of scarring using this is very small. After all, it's only light which really interacts with your skin during this therapy so it only seems logical that it's safer than many treatments that use chemicals.

And while costs vary from region to region, in some areas this treatment starts as low as $100. You'll have to check with dermatologists in your local area to discover your local costs.

Another way to medically remove a wart is to use a chemically based treatment. Typically chemicals used in this approach include salicylic acid formaldehyde, glutaraidehyde, trichlorgacetic acid, cantharidn or podophylin.

## THE GRANDDADDY OF OVER-THE-COUNTER REMEDIES

It's the most common wart removal method of them all, and it's found in many over-the-counter wart removal formulas. The chance that you've heard of it by name, though, might be slim. What is it?

It's **salicylic acid.** Go ahead. Walk into any drugstore to check out its selection of wart removal products. Odds are you'll find at least one product, if not more, with salicylic acid as an active ingredient.

And if you scrutinize these products, you'll find that they can be classified into two large categories. The first are adhesive pads treated with the chemical. The second are bottles of concentrated salicylic acid.

If you've just received a prescription from your doctor for this exact chemical remover, then you're probably confused. What's the difference, you're wondering, between a salicylic acid bought through the pharmacy which requires a prescription and this which I can buy directly off the shelf?

## PRESCRIPTION VS. OVER-THE-COUNTER

In a word: strength. The over-the-counter formulas are for the most part found in a less concentrated form.

They contain only about 17 percent salicylic acid. When you fill your prescription you're getting a product that contains a minimum of 70 percent salicylic acid. Now, there's a difference!

Whichever version you choose, you'll be glad to know that salicylic acid has been called by many as one of "the most effective front-line treatments" for plantar warts, flat warts and the common variety of warts found on the hands as well as the face.

There's no single approach to wart removal that works for all. But in 2004, a comprehensive review of the trials and tests performed on wart removal illustrated that no other method had any advantage over it. Not only that, but this method also carried with it the least amount of possible adverse side effects.

The "do it yourself" wart removal products are usually adequate for the mild and moderate growths. They usually work in cases where there is only a single wart or a small cluster of growths. For the more stubborn, resistant or larger and deeper growths, you may want to visit your doctor to get the more concentrated version of this acid.

## NOW THAT I HAVE THE REMEDY WHAT DO I DO WITH IT?

Before you begin, ensure that you've washed the affected area thoroughly. Keep in mind that the salicylic acid should be applied to the growth only and not to the surrounding skin. Getting some of this acid on the skin itself may result in a burning of the area or at minimum, an irritation.

## DAILY "MAINTENANCE" REQUIRED

Well, yes, you could view it that way. When you choose an over-the-counter removal product, you should be aware up front that this requires a daily commitment on your part. You may discover that it "takes a while" before you even begin to see a diminishing of the wart.

If this frustrates you -- and it's understandable if it does -- some experts offer a tip on how to speed the process along, at least some. Wash the area where the growth is daily. Then instead of drying it with a towel, allow it to "air dry."

As it dries, take an emery board or stone pumice to file the growth away. In this way, you're taking off the dead skin while encouraging the growth of new.

## WHY SALICYLIC ACID?

This specific type of acid belongs to a family called **keratolytics.** This family is believed by the medical community to work on various levels in removing growths. But it's probably the ability to soften the skin which is the most important.

The keratolytic softens the keratin, which is the protein of the skin itself, making it easier for the outer layers to eventual erode and fall off.

## WHAT'S UP
## WITH CANTHARIDIN?

Another chemically based approach to wart removal involves a substance called **cantharidin.** You can find products carrying this ingredient on your drug store shelves under such brand names as Cantharone and Cantharone Plus.

Said to be an effective method of wart removal, cantharidin is actually a chemical derived from the green blister beetle.

Your doctor may prescribe this particular method for you if salicylic acid or freezing the wart doesn't eliminate it. Removal with cantharidin, though, isn't to be found as an over-the-counter remedy. It's a process your physician performs on the wart.

He applies the chemical to the growth, almost as if he were painting it. Then he covers it with a bandage. This is usually a pain-free procedure.

The presence of this chemical causes the skin under the wart to blister. The blistering process lifts the wart away from the skin. When the blister heals and dries, the wart is easily lifted off with the blistered skin.

The only pain may come with the formation of the blisters themselves. If the wart can't be removed following the initial treatment, then a second session may be necessary.

This approach can be very effective on stubborn growths that have resisted being removed by other means. The good news about this treatment is that it leaves no scarring.

# CAUTION:
# WHEN NOT TO USE CANTHARIDIN

This doesn't mean though that cantharidin should be used on everyone. In fact, if you have certain health conditions, such as diabetes or peripheral arterial disease, you shouldn't use this therapy. If you have other circulatory problems you should also refrain from using it. Before undergoing any treatment, make sure the physician knows your health history.

Similarly, while this method is great for many warts, it can't be used on moles or birthmarks. You can't use it on warts that have hair growing from them or warts located on the mucous membranes.

Not only that, but Cantharidin can't be applied on the genital area and it should never be used in conjunction with any other chemical agents.

And . . . that's right, I'm not quite finished with the list of cautions! . . . this particular chemical doesn't come without some adverse side effects.  Once this is applied to your wart, you may feel a tingling, itching or even a burning sensation.  Application of this chemical also leaves the skin very tender for a minimum of two and a maximum of six days on most individuals.

And keep in mind that the application of the chemical agent itself is not painful, but many individuals find the formation of the blister that forms as a consequence of it is!

And if all these cautions weren't enough, here's the kicker!  For all of its benefits, cantharidin is not ap-proved by the U.S. Food and Drug Administration to be used as a removal agent for warts. Since this treatment is not FDA approved and used only when other therapies have failed, it's not widely utilized.


# ELECTROSURGERY? REALLY?

Really.  This does sound rather. . . well, unpleasant.  But it's an effective method chosen by many.  In a nutshell, the two-step procedure uses a high frequency electric current which passes through the wart with the aim of burning it.  The heat of the current destroys the tissue.  And in most cases, it only takes about half an hour to perform.

Before the physician even begins the procedure, he'll thoroughly wash the wart and its adjacent areas.  Then he applies a local anesthetic.  He then heats a needle using an electric device.  The needle is then placed on the wart to burn it.  Specifically this treatment is called **electrodessication**.

Once the wart is burned, the health-care professional can easily removes it with a surgeon's knife.  This portion of the process is called curettage.

Many individuals choose this form of removal because of the immediacy of the results.  One session and the wart is gone.  No second or third visits are necessary.

This two-step therapy is particularly effective on filiform warts located on the face and limbs.  According to the scientific literature, electrosurgery has an overall success rate between 65 and 85 percent.

## Immuno -- What?

More members of the allopathic medical community are investigating the idea that the appearance and growth of warts and the strength of your immune system are related.

The stronger your immune system, the theory states, the less likely it is that you'll develop warts.  Before you dismiss this theory out of hand, you'll recall these growths are after all caused by a virus.  This is certainly a substance your immune system has the potential to fight.

**Immunotherapy** for warts works much like a vaccine which encourages your immune system to fight a specific infectious agent.  The same principle is at work here.

Technically, this treatment's full name is **intralesional wart immunotherapy**.  In a nutshell, the therapy involves a molecule -- usually the Candida antigen -- which is injected directly into the wart with the aim of getting your own natural immune system to fight back.

All of this sounds good in theory, you say, but actually how effective is it in real life?  One study showed that nearly half of those who participated received complete removal of the growths. Another 13 percent had a removal rate of between 75 to 99 percent.

But that's not the end of the results of this study. The statistics just presented only dealt with the warts which were directly injected with the antigen. Now, here's the intriguing portion.

A little more than one-third of the individuals involved also reported that **warts that were not being targeted** for removal at the time of the treatment also disappeared. How many? Between 75 and 99 percent of the growths in other areas.

Another study pitted immunotherapy against cryotherapy. Immunotherapy won. The removal rate of the vaccine-like approach reached 70 percent compared to the 42 percent rate achieved through freezing.

## IT SOUNDS GOOD . . . THERE MUST BE SIDE EFFECTS?

Individuals undergoing the treatment reported a few adverse side effects. Among these included flu-like symptoms lasting no more than 24 hours as well as itching at the site of the injection.

All of this then brings us naturally to the question of vaccines for warts in general. It seems like a logical way to proceed, doesn't it? But remember there are upwards of 80 different types of viruses that cause these warts. It would be rather difficult -- if not downright impossible -- to create a vaccine that would effectively deal with all the possibilities. That's not to say that in the future you won't see such vaccines being developed, perfected and used.

Currently there's only one such vaccine on the market and that's Gardasil. It guards against the virus responsible for genital warts.

# Removing Moles

Moles, much like warts, can be removed in three ways. They can be cut out and then the area stitched, they can be burned, or they can be removed using laser surgery.

## Your Mole and the Laser

I'm discussing laser treatment first because it's the easiest and least scarring method of removing moles -- especially those that don't seem to respond to any other therapy.

With laser treatment, you don't even need to have an anesthetic. According to those who have undergone it, the most prominent sensation reported during this procedure is a slight tingling like they were being snapped with a rubber band.

The procedure itself is pretty simple. The laser is aimed at the mole. The effect of this is the sealing of blood vessels. The laser also evaporates the tissue. And voila! The mole is gone.

Because of this it's not necessary to do any cutting on the growth or the skin. And the best part is that the odds of scarring are slim to none!

After the laser surgery has been completed, you'll notice a scab forms in the area where the mole had been. It'll take about two weeks to fall off on its own. The only noticeable sign of the procedure could be a slight redness where the scab had formed, but don't worry. It's not a permanent coloration. The redness goes away after a period of time.

## MOLE REMOVAL:
## CUT IT OUT!

One method of removing moles is what emedicinehealth.com calls "simple cutting without stitches." This description is perfect. The surgeon uses a scalpel to shave the mole so it's not raised above skin level. Ideally, he's striving to bring the growth down below the level of your skin.

Once he's accomplished that then he will use an electrical instrument that cauterizes or burns the area to remove the remainder of the mole. In some cases, he may place a solution on the area as well in order to stop the potential blood flow.

Then a topical antibiotic is applied to the wound and it's covered with a bandage. Before you leave his office, the surgeon supplies you with detailed instructions for caring for the area.

# Cut It Out With Stitches

Sometimes your surgeon opts to remove a mole using excision  cutting and then stitching the area. For the most part, this particular technique is used on those moles which have a darker color to them or those that are flat or are both dark and flat!

The surgeon maps out the mole, sterilizes it and then finally numbs it. Using a scalpel, he cuts the mole and the border surrounding it. The actual size of this border depends on the possibility that it's either pre-cancerous or cancerous.

Once the mole has been cut out, he'll then stitch up the area. The depth of the stitches depend on the depth of the cut. If they're placed deep, then your physician uses stitches that are easily absorbed by your body. This means they won't need removal.

However, in some cases, he'll stitch just the upper surface. He won't use absorbing stitches for this. And you'll need to make a return visit to get them removed.

In either case, you'll need to make sure you apply a layer of antibiotic salve as well as a bandage to the wound. Don't place the salve on though until you have thoroughly cleansed the wound. The minimum you should clean it is once a day. It would be ideal if you could clean it twice a day.

It doesn't take much to keep the area clean. Water will do nicely. If you want something more, try a solution of diluted hydrogen peroxide.

Continue doing this until the wound is healed.

## THE HEALING PERIOD

Contrary to what your mother or your grandmother may have told you (sorry Mom!), a wound does not necessarily need to be open to the air to heal properly. Recent scientific studies have been performed comparing the two methods. Surprisingly some wounds heal even faster when they're properly bandaged with antibiotic cream.

Today, many individuals boast that they get a quicker healing response when they use **vitamin E on the wound**. This may work for some people, but not necessarily for all. Believe it or not, vitamin E may actually retard the rate of healing. Use it if you think it may help you or if you've already used it in similar instances with success.

If you don't have an antibiotic cream handy like Neosporin or Polysporin, try using petroleum jelly. Many people discover this works just as well to aid healing.

# WHAT'S NEXT?

Unless you need stitches removed, there's no need to see your doctor again about the excision. But even here, there are exceptions. For example, your physician's office may contact you about the results of lab reports if tests are performed on the growths.

Should an infection develop, then it's important you go back to the attending physician for possible treatment. Chances increase of contracting an infection if you allow the wound to get dirty.

So how do you know when to visit your doctor? There are several telltale symptoms that a second visit is needed. The first is the appearance of any excessive discharge, either pus or blood from the wound. This is especially true if the fluid is foul-smelling.

If you develop a fever greater than 100° F. This holds true for any adult in fact. In a child though, seek medical care when his or her temperature rises to 101°F.

If you experience lingering pain that can't be relieved with over-the-counter pain relievers, then seek medical attention.

And oh yes, about those stitches. If you have any on the face, you'll probably see your doctor again with four to seven days. If they are anywhere else on your body, you won't need to visit for at least eight or as many days as 21. The variance in time depends on the sutures as well as your doctor's opinion.

Most people don't go through the bother to remove moles. But if you do decide this is an option for you, remember that just about everything comes with some type of risk, and this is no different.

# Risks With Mole Removal

Some people develop an infection following mole removal; others will inevitably scar (of course, that's not to say you will!). Still others discover they have an allergy to the anesthetic.

One way to reduce these risks is to carefully choose your dermatologist or surgeon. Make sure he or she has the right skills needed for this treatment.

## Get this tag off of me, please! Removing skin tags

Actually, that's really quite unnecessary. Because skin tags are benign, they don't require removal. In fact, a doctor won't even recommend their removal unless it's absolutely necessary.

There are only two reasons getting rid of these tags are deemed essential. The first is that they are continually being irritated and getting sore through the rubbing of clothing or jewelry. In this case they become a blatant case of discomfort. The second cause for removal is if their appearance in a certain locale should constitute a cosmetic problem.

Individuals with skin tags ask a common question: If I get my tags removed, will even more appear? There is no medical evidence of such an event. They won't "seed" or spread if you get your existing growth taken off.

Having said that, you should also know that some persons are just more prone to the growth of these tags. For these individuals, whether they had the tags removed is irrelevant. More tags may just automatically appear for them. But this is definitely independent of the removal of any prior to this -- that is, there is no cause and effect involved.

The procedures for the actually removal of skin tags are really quite simple. The most common methods are simply to burn it off with an electric needle (electrosurgery) or to cut it off with a scalpel or scissors.

The healing process that follows typically takes no longer than a week and at most two weeks. But this procedure doesn't guarantee that they won't return. Because they just might.

More traditionally, skin tags were removed by what I call a "choking" method -- even though that's not a very scientific term. The physician ties the base of the tag -- that tiny stalk -- very tightly using a piece of cotton.

## WAITING FOR MORE?

Most people do. But that's really all there is do it. And then the individual affected and the physician wait. What exactly do they wait for?

They wait for the tag to simply fall off. And it eventually does. The tying of the stalk stops the blood flow to the tag itself and it just dies a natural death, so to speak. And now you see why I like to call it "the choking" approach.

The problem with this method, though, is that in some cases it leads to the creation of an infection.

Some physicians use instead **cryotherapy**. If you recall, that's the process of "deep freezing" the growth. I've discussed it in length when referring to moles.

When this method is applied to skin tags, its approach can be quite different. In this case, the liquid nitrogen is held in a paper drinking cup. Only a small amount of the nitrogen is poured in there. Your physician dips a hemostat, a non-toothed forceps or even a needle-holder into the liquid for about fifteen seconds.

Your physician then gently grasps the stalk with one of these instruments for approximately 10 seconds. The surrounding skin must not come in contact with the freezing agent.

Up to ten lesions can be treated with a single dip into the liquid. You'll only fee a mild sting, if that, while he's performing this task and if done carefully, it won't damage or irritate the surrounding skin.

No type of dressing is necessary to cover these areas once the cryotherapy is complete. Within seven (7) to ten (10) days the tags will simply fall off!

Many physicians say this specific method is of great use when removing pedunculated tags which hang on the eyelids.

## OVER-THE-COUNTER REMEDIES FOR SKIN TAG REMOVAL

If there's a health condition that needs "fixing", there's an over-the-counter remedy that promises to fix it.

In this case you can purchase several topical products, all claiming to be able to remove the unwanted growths in a matter of minutes. But most of all, they all claim to be able to remove them for less -- much less -- than the cost of a trip to your dermatologist.

Of these, the most popular one seems to be a product called DermaTend. This particular formula is quite versatile, claiming it can also remove warts and moles.

Despite it being a multi-step process, it's fairly easy to use. The steps, by the way are spread out over a period of one week.

The first action you're required to take is to exfoliate and pumice the area. Yes, that's right. The idea behind this is to make the area tender. You'll then put hydrogen peroxide on the area, followed by the DermaTend.

Once you've done that, you'll cover the area firmly with a bandage and leave it on overnight. When you take it off, examine the tag.  Chances are it has formed a scab. Place a hot cloth to the scab in order to soften and open the pores.

Now, you're ready to reapply the DermaTend.  Do this every day for about a week.  After that the tag should be gone.

## HERE'S ANOTHER OTC PRODUCT

The second over-the-counter product is called **Heal Skin Tags.**  This particular formula is composed more of herbs and essential oils rather than anything harsh and artificial.  You only need to apply a few drops of this natural formula on the growth.  The tag should fall off in two (2) to six (6) weeks.

This product works by actually drawing out the tags from their roots.  In this way, they'll either flake away or fall out.

Well, why yes, there still are several ways you can safely remove skin tags without resorting to expensive visits to your doctor.  The following two chapters talk about just that.  They both deal with various "home remedies" for removal of moles, warts and skin tags, as well as herbal and nutritional methods.

From among all of these, hopefully you can find several from which to choose that will help you get rid of that skin growth that's making you feel so uncomfortable.

# Chapter 3:
## Herbal And Nutritional Treatments for Removal of Moles, Warts and Skin Tags

Yes, I know. And I know because I've been where you're sitting right now. You're back in the waiting room of your physician's office. You're frustrated, you're definitely confused, and you may even be angry.

And you have every right to be. Why? Because you've just spent a lot of money for your mole, wart or skin tag removal and well . . . nothing got removed. Nothing that is but money from your pocket. And you still have your growths.

Before you go back to your doctor for him to do whatever you had in mind, why not look over some of these herbal and nutritional treatments for removal of the growths.

In the following remedies just about all the remedies listed for removal of warts works for moles as well. I've added a few that are designated as effective mole removal, but these two can be used with great results on warts. You'll see that I tried to put some sense of classification to them. Hopefully that should help you.

I know what worked for me -- and I've researched what worked for others. And this is the result of those months of research, emails and general discussions on the discussion boards.

It's my sincerest hope that you find your own unique and special approach to dealing with your skin problem.

# Removing Warts With Herbs

## Calendula

This herb works because it has amazing anti-viral potential. Make a very strong tea using calendula as the base. You'll want to steep two (2) tablespoons of the herb for about 20 minutes in a cup of boiling water.

You can use this herb in two distinct ways to help the removal of the wart. Of course, you'll want to apply the tea to the wart itself. But you may also want to drink the tea as well. Just prepare it to your particular taste.

## Tea tree

Using a cotton swap, apply the undiluted tea tree essential oil to the wart. Do this several times throughout the day until the wart disappears.

## Echinacea

This particular remedy deals more with the root cause of the wart more than the physical removal of the object itself. Consume a dropper full of tincture of Echinacea three (3) times daily for ten (10) days.

Echinacea is well known as an herb that boosts your immune system. The goal is to increase the health of your immune system, which fights germs as well as virus invasions and those warts that trouble you will have a harder time making a home on your skin.

# Aloe Vera

Believe it or not, this particular herb has been used for hundreds of years as a wart remover. You can buy aloe cream to apply to the wart. If you have a plant, though, don't be the least bit afraid to apply the sap of the plant to directly to the wart.

The application of the plant also eases any discomfort you may experience because of the wart.

# Chickweed

Originally native to Europe, the chickweed plant can now be found throughout the United States and most of North America -- as a common weed.

But the "weed" can be very valuable to you in your attempt to remove your wart or mole. This particular "recipe" calls for you to crush the chickweed. Place this crushed material on the mole or wart, and then cover it with a bandage. Keep on doing this until the mole disappears.

Of course you can always find fresh chickweed, but you can easily discover this herb at your local health store. It can be found in capsule or liquid extract tincture as well as an ointment and oils. And of course, you should be able to find it anywhere in a dried herb form.

# Got Dandelion Milk?

Never knew dandelions had milk? You probably knew it existed, but you may not have known its name. This is the milky sap that's inside the stem of the plant. Simply apply the milk to the wart several times daily.

Other people have claimed that the root of the dandelion is also an effective wart remover. The root should be rubbed on the wart itself until the "juices" from it come out and cover the wart completely.

Do this routine at least twice and preferably three times a day until the wart disappears.


## Extract of Milkweed Herb

While this may sound something that one of the witches in Shakespeare's **Macbeth** uses, it really is a very effective herbal remedy.  You can try it if you don't believe me.  Just apply the extract of this herb to the wart. Leave it on overnight.

Perform this routine for one week.  The mole will disappear shortly after this.

## Get on the A-List
## The Vitamin A List, That Is!

For this remedy, you'll need vitamin A supplements containing 25,000 IU of the supplement preferably from fish oil or fish liver oil.  Break the capsule open and squeeze some of the liquid from inside onto the wart.  Be sure to gently rub it in thoroughly.  Do this daily.

Be aware though that this is not a quick fix.  According to Dr. Robert Garry, associate professor of microbiology and immunology at Tulane University School of Medicine, you'll have to perform this routine for several months at least.

But if you're patient, the wart will disappear.  The actual length of time it takes may depend, in part, on the type of wart.  If the wart is on a child, it may leave in as little as one month.  More typically, though, this treatment takes at least two and usually four months to entirely eliminate the wart.

One of Dr. Garry's patients had more than 200 warts on her hand.  He treated all of them, he explains, using only this vitamin A technique.  Of those, all but one was eventually eliminated.  Not bad odds!

## Vitamin C

Vitamin C has been a superhero among vitamins for decades now because of its ability to ward off colds and viruses.  It's a great nutrient to use when you need to improve your immune system. It's not surprising, then, to see it pop up in some form as an effective wart removal method.

Crush several vitamin C tablets.  Apply these directly to the wart.  Cover the growth with an adhesive bandage.  Do this for several days.  You'll discover that the growth will disappear in a relatively short period of time.

This remedy comes with a doctor recommendation, just like the vitamin A suggestion.  Dr. Jeffrey Bland, who studied at the Linus Pauling Institute in Menlo Park, California, thinks that this removal process works because the "high acidity of ascorbic acid [of the vitamin C] can kill the wart-producing virus."

However, think twice before you actually proceed with this removal suggestion.  Vitamin C -- in its ascorbic acid form -- may irritate your skin.  You should try to cover only the wart with the crushed supplement.

## HERBAL HELP IN SKIN TAG REMOVAL

## Tea Tree Oil

Now it's time to tap into the plant known as Melaleuca Alternifolia?  What?  You've never heard of it. Not surprising.  You probably know the essential oil that is extracted from the plant though: tea tree oil.

While it's legendary in its healing powers, did you realize that this oil can also aid in the removal of those unsightly and unwanted skin tags. I can hear you know: Tell me more.

It's quite simple actually. First, make sure the affected area is clean by washing it thoroughly. Dry the area well. Then dip a small cotton ball in water just enough to dampen it. Place two to three drops of tea tree oil on the cotton. Rub it on your tags.

Repeat this procedure three times a day until the growth vanishes. And vanish it will!

## Bloodroot?

Sounds serious. Bloodroot is a type of herb which grows in the north central part of the United States and Canda. It's possibly called bloodroot (just because I know you're wondering about its name!) from the color of the juice harvested from its root. Yep, it's red.

And according to Dr. Weil, the well-known natural health physician, it is serious. The bloodroot herb is a strong herb, so strong in fact that it's poisonous if taken internally.

So I recommend this remedy with a bit of trepidation and much caution. But this herb, scientifically known as Sanguinaria Canadensis, has a long and illustrious history as a remarkable remover of skin growths, especially skin tags.

You can buy this herb at many health food and vitamin stores or you can easily find it on line. You can buy it in powdered form or as a paste.

You'll want to apply it to the growth and cover it with a bandage.

# Vitamin E
# To The Rescue

If you already have vitamin E capsules on hand then you're in business. You can start this remedy any time (following your double checking with your physician of course).

Start off by opening up a vitamin E capsule and carefully placing some of it on a bandage. Apply the bandage to the skin tag and leave it in place until it loosens from your skin. Take the bandage off. If the tag is falling off the session, it has been a success.

If the tag isn't quite ready to fall off, repeat the entire process. Keep repeating until the flap falls off.

Many people have found that their skin tags disappear nearly magically when they take 200 mcg of chromium picolinate twice daily. A trace element, you can find it in just about any health food or vitamin shop.

Another mineral with a similar name, chromium poly-nicotinate, is also effective as a tag remover. This particular mineral though takes quite a while to work -- sometimes as long as three months. And beware, because once you stop taking the chromium polynicotinate the tags return.

If you're considering taking this mineral, you should consult your physician beforehand to get his advice and input as well.

If you're concerned that this chapter is winding down, you may be wondering if there aren't any more remedies. Well, then you'll be interested in the following chapter.

The suggestions for removing warts, moles and skin tags found in this chapter include foods and other diverse items you probably already have in your home.  And yes, the infamous "duct tape" removal process is included in this chapter.  Why that one removal method alone is worth reading the entire chaper, isn't it

# Chapter 4:
## "Home Remedies" for Removal

Open that kitchen pantry.  Take a good look inside. Don't hesitate to walk through you home.  Go ahead.  Your bathroom too.  Unless you're totally obsessed with finding 101 ways to remove moles, warts and skin tags (and who knows you just might be!), you may be overlooking a myriad of options.

Many of the remedies suggested in this chapter have been handed down from generation to generation.  And many of the remedies come complete with a large contingent of individuals who swear by their successful use. And yes, many of the remedies really work surprisingly well.

Others may not work for you, but your neighbor may use nothing else but this one off-the-kitchen-shelf item.

That's what this chapter is all about: common, everyday items that could possibly save you yet another trip to the doctor for a skin growth removal session.

Think it's impossible that something like garlic or apple cider vinegar can remove warts?  Finding it difficult to believe that your mole may come off with just prolonged exposure to grapefruit juice?  As wild as it may sound . . . . as far-fetched as you may believe it to be . . . others who have suffered have discovered relief.

Below you'll find just a few of the home remedies people use every day for the removal of warts, moles and skin tags.

# Warts

## Apple Cider Vinegar

Apple cider vinegar has been proffered as a home remedy for every type of ailment imaginable from weight loss to heartburn . . . and now for the safe removal of warts. This remedy is so safe that it's been used on children as young as six years of age.

All you need to get started is:

- Apple Cider Vinegar
- Petroleum Jelly
- Cotton balls
- Waterproof medical tape -- about 1 inch thick

Each evening before you head off to bed, coat the skin surrounding the wart with petroleum jelly. Then take just about a "capful" of the apple cider vinegar. Dip the cotton ball into the vinegar. You'll then place this vinegar-soaked cotton ball onto the wart. Wrap your hand with the waterproof tape.

Every morning when you wake up, take the bandage off. Start the procedure all over again the following evening before you go to bed. Individuals say this is a natural -- if slow -- process. For some, it takes up to one month to get rid of the wart. Others reported it only took a matter of five (5) days for the wart to drop off!

## Castor Oil

This remedy is really quite simple. Just rub a bit of castor oil into and around your wart. Do this two (2) or three (3) times a day. After you rub the oil in, cover it with a bandage to keep the castor oil on and around the site as much as possible.

Consider adding some baking soda to this. Some people say it helps to improve the effectiveness of the castor oil. Applied before you go to bed at night, you'll discover your unsightly blemish will be gone before you know it.

## Would you like potatoes with that wart?

Please notice I didn't say fries! This suggestion for wart removal involves taking a raw potato and cutting it in half. Rub the inside of the potato on the wart. You'll have to rub the raw potato on the wart several times a day. Continue this routine for a minimum of two weeks.

If you're truly ambitious, take the second half of the potato, the unused portion, and plant it outside for the start of a tasty vegetable garden.

There's also a version of this remedy that suggests you grate a potato, place it on your wart or mole and cover it with a bandage. Allow the potato to remain on the mole for about 6 days. When you notice the bandage is falling off naturally, remove the bandage and the grated potato. Your wart or mole should be nearly ready to fall off.

The key to this version, it's said, is the decomposition of the potato which activates the wart removal process. (I'm not so sure I want to live with decomposing hash browns!)

## Lemon Tree Very Pretty . . . And the Juice Removes Warts?

Apparently so, according to some individuals who have suffered for years with warts. But don't use it alone. Place several slices of lemon in apple cider. Add salt to this. Allow this to stand for two weeks. Then you rub the lemon slices onto the wart.

# Wait . . . You Want Me To Buy A *Banana* For My Plantar Wart?

*I could have sworn you said to buy a banana for my plantar wart. You did say buy a banana as part of a home remedy didn't you?*

*Are you sure you're feeling up to par?*

*Well, my plantar wart is really hurting. So I guess I don't have much to lose. Oh, you're sure this is going to work?*

*Take a piece of ripe banana peel. Apply the pulp side -- the inside of the peel -- onto the wart. Now, cover this with a bandage. The only time you'll remove this is when you're taking a bath. Apply a peel to this wart every day. While it works, those who have used it say it takes several months before the wart disappears.*

## Why Not Garlic?

Good question. It keeps vampires at bay, why wouldn't it make warts disappear? Actually, garlic is a logical choice due to its natural anti-viral properties. This means it's a great plant at fighting viruses. Herbalists everywhere routinely urge individuals to eat fresh garlic in order to boost their immune systems. So it only makes sense that it takes those nasty skin growths away as well.

This remedy takes a little more preparation than simply taking some juice and bandaging the area. When you use this remedy, you'll want to make a poultice of crushed raw garlic.

Before applying this poultice, be sure to protect the other areas of your skin using vitamin E. Simply take a capsule, prick it open with a pin, and apply it to the skin surrounding your wart. Now you're ready to allow the garlic to do its thing!

Take the garlic that you've just mashed and apply it directly to the wart. Cover the wart with a bandage for a full 24 hours. At the end of this period you can remove it.

What you'll discover under the bandage is a formation of a blister. Within a week of this, the wart should just fall off.

## Grab That Grapefruit!

Quick! Grab that grapefruit. Squeeze it to extract its juices. Now apply this juice to your moles or warts. Do this several times throughout the day. Continue doing this for about a month and your growth should be gone.

## Onion Juice
## Is Always An Option

Chop a raw onion. Cover it with salt. Allow it to sit undisturbed overnight. Then take that juice and apply it to your wart at least three times a day until it disappears.

## Don't Apply The Cabbage . . .
## Eat It!

That's right!  This specific home remedy tells you to eat as much cabbage as you can stand.  You can even eat it in the form of coleslaw if you like.  If you do this for nearly two weeks, you should see the warts disappear!

This sounds like a great remedy.  Cabbage is also a great vegetable high in all sorts of antioxidants and vitamins and minerals.  What have you got to lose in eating it?  You can only be improving your health!

## Grab a Pineapple . . .

And apply it. Don't eat it!  Confused yet by all these remedies?

The next time you're at the grocery store, pick up a fresh pineapple from the produce section.  It just may be the answer to your wart problem.

Take a slice of this fresh pineapple and rub in into your wart before you go to sleep at night.  Be sure to use a circular motion when doing this.  Also, you'll want to make sure you rub it in for a minimum of five minutes at a time.  Allow it to soak into your wart overnight.  This means you can't wash the area before you go to bed.

When you wake up the next morning, though, thoroughly wash the pineapple-soaked area.  Continue to do this for two days.  You'll notice no change in the wart that second day, but you're done treating it with the pineapple.

The following day, examine the wart, even though you have done nothing to it.  You'll probably notice that the root of the growth is beginning to change colors.  Keep your eye on the blemish.  Within two weeks it should be gone.

If, after that period, the wart hasn't disappeared, use the fresh pineapple rub again for another two nights -- and two nights only.  Continue this routine until the wart falls off.

## Grapefruit Seed Extract

Apply a drop of grapefruit seed extract to the wart. Cover it with a bandage.  Do this several times a day always using a clean bandage when you cover the growth.  Individuals who have used this method say your wart should fall off within four weeks or less.

## Flaxseed:
## Oil and Ground

Those who have used this remedy highly recommend it. It uses flaxseed in two forms, both ground and its oil.

You'll take the two forms of flaxseed and mix them to-gether in a poultice, along with just a small quantity of raw honey.  Apply this to the wart itself.  Change the poultice on a daily basis.  Before you know it, your wart has disappeared.

Don't know what a poultice is or how to make one? You're not alone. It's not something most people know about readily, unless their herbalists showed them how to make one or they learned it from their grandparents.

## The Poultice In Action

A poultice has been used for centuries (and probably longer) for a variety of reasons and with all types of herbs and natural substances to draw infection from an area and speed the healing of wounds.

They work, so herbalists and alternative healers say, by increasing the blood flow to an area, as well as relaxing tense muscles, and soothing tissues.

You'll want to get a piece of gauze, muslin, linen or even just plain white cotton. Lay this out on a flat surface. You won't need a large amount, just enough to cover the affected area. Since the affected area in this case is a wart, the poultice will be rather small.

Spread the ground flaxseed and oil mixture on the material. Clean the affected area where the wart is thoroughly, preferably using hydrogen peroxide.

But wait you're not quite done yet. Use a safety pin to hold the poultice in place and cover with a towel (this prevents the flaxseed from getting on your clothes, depending on the location of your wart). If you'd like, use a hot water bottle to keep the poultice warm.

## NOW, I'VE HEARD IT ALL! DUCT TAPE TO THE RESCUE!

Duct tape is used for just about everything -- even NASCAR uses it to temporarily repair a race car so it can stay competitive in a race. But, could duct tape also be the perfect treatment for the removal of warts?

Well, okay, so perhaps it's not the "perfect" treatment. But is it really an effective treatment? It seems that it just might be. There are many individuals who swear by this remedy, and now, just when you think it can't get much stranger than this, another voice in favor of duct tape emerges.

This one comes from a study conducted by Dr. Dean Focht and reported in the *Archives of Pediatrics and Adolescent* medicine. The study discovered that wart removal by duct tape is actually a more effective wart removal method than cryotherapy.

Two different approaches have been reported, each producing good results.

In the first procedure, you apply the tape to the wart keeping it on for a minimum of six days. When you remove the tape, you'll want to soak the wart and eventually pare it down with an emery board.

In the second method, apply over-the-counter salicylic acid wart remover liquid to the wart before you go to bed at night. Allow the site to dry for about a minute to a minute and a half. Then apply the duct tape. Ensure the area is completely covered. The next morning, you can remove the tape.

You'll repeat these steps every night until the wart tissue has disappeared. Each time you remove the tape, you are also removing a bit of the wart tissue.

Of course, there are a few disadvantages to this treatment method. First, with the repeated application of duct tape, the area can turn very red and soggy where the tape sits all night.

If you find this happening, simply skip a day or two of treatment. Allow the skin around the wart to breathe and dry. After the skin is restored to normal, you can continue with the method.
This is also a very gradual process. You're not going to see results over night. Finally, treating your wart with duct tape requires a commitment on your part. It's not a "quick fix" which you slap on the unsightly growth and -- viola -- the following morning it's gone, as if it never existed.

But if you want to avoid the pain of some of the more conventional treatments and are willing to dedicate some time and effort, you may have just discovered a great wart removal treatment.

# How Does Duct Tape Work?

That's an excellent question. Doctors are still posing that very same question. Right now two theories exist to answer that. The first is that as you remove the tape from the wart, you're also removing some of its dead skin. This gradually gets rid of the wart virus itself.

The second theory on why duct tape works suggests that when the growth is covered like that, the immune system is activated to attack the virus.

Whichever the scenario -- or possibly one not even posed yet -- it really does seem to work.

# Got Walnuts?
# No, Not Just Any Walnuts.

These have got to be ripe. And, oh yes, still in the shell. Go ahead. Go get some, I'll wait. Find some? Good.

Now, your next step is to carefully cut the outer shell. Obviously, you'll need a fairly tough utensil for this job. You need to cut enough of the shell that the "walnut juice" oozes out. This is exactly what you're after - the "juice".

Rub the liquid from the walnut shell onto the mole. Be prepared to experience a tingling sensation if you have sensitive skin. Some individuals may even notice their skin darkening in the areas where the juice has been rubbed. Don't panic. This is normal. In fact, it's a sign that the juice is actually doing exactly what it's sup-posed be doing.

When the area turns dark, it can be interpreted that the removal process has begun! And don't worry, the change in skin tone isn't permanent. It isn't even going to linger overnight. Within a couple of hours, your skin will return to its normal color.

Continue this routine until the mole falls off.

## Pass The Honey, Sweetie!

If you have nothing else at home and you need an im-
mediate fix, look to the healing power of honey. It's a
product just about everyone has in the house.

The fix itself is easy enough. Simply apply the honey on
the mole several times throughout the day. This alone
will make the mole disappear completely. And it won't
take much more than several applications.

Yeah, it may be a bit sticky, but who I am to argue with
success -- even sticky success?

## Got Children's Chalk?

Yes, that's what I asked. But it has to be white chalk
(and no, I'm not sure why it has to be white). But if you
apply this on your mole regularly, it will lighten the
growth and eventually remove it altogether.

## SKIN TAG REMOVAL

If you recall, I said that skin tags were just different
enough growths from warts and moles that they needed
to be approached from just a slightly different angle.
Below are some of the different angles you might want
to try!

## Why Not Apple Cider Vinegar, Again?

Apple cider vinegar comes in handy once more. Is there
any reason to think that it couldn't handle skin tag removal? If you thought otherwise, here's your chance to
discover just how effective this legendary remedy can
be.

According to those who have used it successfully, all you need to do is rub the vinegar onto the tag using a cotton ball three times a day. There's no need to cover the tag or give it any type of special treatment. It should fall off naturally in ten (10) to fifteen (15) days.

I feel obligated to tell you this next super simple method of removing skin tags. Be aware that while this method is praised by many, it's also an approach many physicians advice against. What is it?

## Snip It Off!

Simple snipping of the skin tag stalk. For those who have tried it, the general consensus seems to be the use of a pair of nail clippers, thoroughly cleaned and sterilized. The clippers are suggested because they are easier to handle than most other cutting utensils.

To ensure sterilization of your "instrument" of choice, you'll want to plunge the clippings into boiling water for a minimum of 5 minutes. When you take the clippers out, use a cotton ball as well as some rubbing alcohol to ensure its sterilization.

Make sure the clippers are cool before using them. Now, that you've done all that, you only need to snip the tag off at the bottom of the stalk.

It may sting some, but in general it won't be painful. You'll want to have some cotton balls as well as peroxide nearby just on the chance you bleed some. And if you don't bleed you can still use these two items to cleanse the area thoroughly.

You'll want to follow this cleansing procedure with the application of an ointment, like Neopsorin, to the area. This will help to prevent the possibility of an infection forming here.

# Choking The Tag

If you'll recall, "choking" the tag is one of the options your doctor has at his disposal when you visit him in search of eliminating these guys. It seems quite normal that this method has already been tried in the privacy of people's home -- with success.

In fact, according to some accounts, the doctors of to-day actually received the ideas by watching their grandmothers of yore treating the tags at home. In any event, many individuals have done this successfully on their own.

Simply take some thread and tie it around the base of the tag tightly -- really tightly. The goal here is to elim-inate the blood supply from the stalk to the flap itself. Once the tag isn't receiving the nutrient supply it needs, it will then die and drop off on its own.

This is actually a much safer approach than snipping the tags with the clippers. There is virtually no risk of any additional infection.

# Suffocate Them!
# Nail Polish Is The Way!

I suppose that's how you would describe the next ap-proach. If a substance can't breathe, it would have a hard time living. And if you cover the tag with fingernail polish, then the growth won't be able to breathe. Do this two to three times daily, until the flap falls off, which usually takes only several days.

# Duct Tape Them!

Just like there's a remedy for wart removal involving the all-around marvel duct tape, there's a similar remedy using duct tape for skin tags. Simply cover the growth with a piece of duct tape. Keep it on until you notice the tape beginning to loosen up a bit. Remove it. Examine the tag. If it's not ready to fall off then place another piece of tape on it and repeat the same procedure. Do this until the tag falls off.

## Hash Browns For Tag Removal?

Well, not quite. But it comes really close. One method that many individuals take to the bank (or is that to the breakfast table?) is to place some raw grated potato on the tag itself. (Yes, I couldn't make this up if I tried!) Cover it with a bandage and allow it to remain on your tag for five to seven days.

Now, as much as you may want to, don't change the potato gratings on a daily basis. The key to this remedy's success lies in the decomposition of the potato. As the decomposition occurs, the tag will eventually fall off.

## An Aspirin A Day . . .

Who knew the simple aspirin was so versatile? If you're squeamish about having decomposing potato on your skin, then you may want to try the **aspirin cure.** In this home remedy, you dissolve a tablet of aspirin, making a paste with it.

You apply this aspirin paste to the skin tag. Then cover this with a bandage. Repeat this method twice a day until the tag falls off.

## Garlic and Vitamin E
## Tag Team the Skin Tags

And why not? Why not use the best of the nutritional and the best of the herbal remedies to tag team your skin tags into oblivion . . . and beyond!

Vitamin E, by the way, has been used for several decades for improving your skin with marvelous results. So it's no surprise that it's a key ingredient in this home remedy.

You'll want to use vitamin E oil -- which can easily be obtained through the act of breaking open an E capsule. Take the liquid from this capsule and mix it with garlic to make a paste. Cover the area. You may have to repeat the procedure depending the size and aggressiveness of the growth. But this one-two combination will knock out the skin tag!

## Baking Soda and Castor Oil

Similarly many individuals use a paste of baking soda and castor oil as the basis of a home skin tag removal method. Place this paste on the tag, place a bandage over it and allow it to soak into the tag overnight. Continue to do this until the tag falls off.

## Crushed Garlic?

Well, if you can stand the smell, you can rub crushed garlic on the tag three to four times daily. This promotes the gradual disintegration of the growth. You may have to do it for several weeks before anything happens though.

## Less Malodorous But Not By Much

Don't like the smell of garlic on your body? There's probably a good reason no one has developed a garlic-based perfume or cologne. If that's just too much for you to stand, consider the onion. Okay, so it's not much better.

Specifically, think about onion juice. Chop an onion. Place some salt on the chopped pieces to draw out the juice. Do this daily until your tag falls off.

You can also follow the same steps using cauliflower juice. Apply the juice to the tag in a circular motion. Do this several times a day and before you know it, the tag is gone.

There are enough home remedies here that you can certainly find some removal method with your name on it. Of course, I've mentioned this in several areas of the book. But there's still one more avenue we need to go down. And that's the possibility of actually preventing the warts from forming to begin with.

Impossible you say? Perhaps. Definitely there are no promises. But the following chapter shows you how you can at least reduce your risk of getting these growths.

# Chapter 5:
## Prevention: Easier Than You Think

**M**ost people seldom talk about preventing the appearances of warts. And in many cases, you simply can't. The viruses causing these pesky things must be ubiquitous -- they're just about everywhere!

But the very fact that they are caused by one of 80-some viruses should mean that you should theoretically be able to avoid coming in contact or contracting the virus to begin with.

Hmm. Now, we're starting to make sense. No, there's not a long list of do's and don'ts involved in this. (In fact, you'll be amazed at how short a chapter this is!) But there are several guiding instructions you should follow if you want that wart that just fell off because you treated it with duct tape to be your last wart.

Now, I have your attention.

## SO WHERE DO I START?

With the basics. And that means reviewing one more time -- bear with me -- that warts are caused by viruses. Here's the logic in my thinking and the thinking of many health care professionals. If you avoid the virus, you can avoid the wart. Yes, it's that simple.

Well, yes, it's that simple on paper. And the following tips will make it just a little bit easier for you to find practical ways of avoiding the virus in your daily routine.

# SHOES

We're starting with your shoes. The virus that brings you the tenacious and persistent plantar wart thrives in a moist -- very moist, -- environment. This observation and advice comes from Suzanne M. Levine, D.P.M. This podiatrist suggests that when you're at swimming pools, health clubs and even in locker rooms don't ever go barefoot. At the very least wear thongs -- you know, what everyone these days calls "flip flops."

If you follow this advice (which is probably exactly what Mom told you when you went away to camp every year!), then you can avoid your feet coming into contact with the virus.

The corollary to this is to not go barefoot. I know that bothers the dreamers and more sentimental types reading this. But going barefoot outside only screams at the world to bring on the little cuts or cracks which may allow the entrance of the virus-causing warts.

In addition to wearing shoes, change your shoes as often as you can. Again, it all comes down to the idea that the offending virus loves a moist home. If you trade off shoes -- wear one pair on Monday, Wednesday and Friday and pair two on Tuesday, Thursday and Saturday -- you decrease your risk of your shoes getting exceedingly moist.

## Become Mr. or Mrs. Clean

There's no need to become obsessed with cleanliness, but you might be able to avoid the virus by taking tighter precautions when it comes to cleaning. This is especially true when it comes to the bath and shower. It doesn't take much more than a little discipline and a little bit of elbow grease to keep your bathroom clean.

## Use A Shower In A Gym?

Then you may want to be a little more vigilant than you've been in the past.  While I don't want to use the word "obsessed" here either, you may want to consider washing the shower at your gym or health club even before you jump in with a germ-killing product.

## Don't Touch Your Current Warts

This advice is especially true of plantar warts, because they are the most contagious of all the skin growths but it's good general advice. Warts spread so easily.  It's far too easy to touch the bottom of your foot without even knowing it and then realize you have a small cut on your finger.  Oops.

# Operation:  Protection Cuticle

No, it's not the latest top secret project of the Pentagon. It's your new mantra.  Warts love to enter into your system through a cut or some other type of opening around your cuticle.

If you'll recall, when the wart is located under or on the nail area it's called a periungual wart.  Not only are these painful, but they're also among the most difficult warts to remove.

These warts get their genesis when the virus enters through a cut or opening around your cuticle, that area that surrounds your nail.  It only makes sense then that the less cracks and cuts you get in this area, the greater your chances are of avoiding the development of growths.

If you have the habit of biting your nails, make a concerted effort to stop.  This way you know that the virus isn't entering through any bites or cuts caused by your "nibbling" at your nails.

If your child is prone to warts around the nails, ensure that their nails are always trimmed.  This also discourages them from biting their nails.

Another effective preventive approach and one overlooked for this purpose is frequent washing of the hands.  If you have accidentally touched a wart, then the more often you wash your hands, the less likely it is that the virus is linger on your fingers and hands. And of course, it just helps to keep your health overall in better shape.

And while we're on the topic of nail clipping, you need to add one more habit to your list.  Don't loan your nail clippers or files out.  Actually there are a few lessons here.  That's number one.  They just may come back with some virus on them.

If you're the one with the warts around the fingernail, then you'd be better off to buy a second pair for yourself.  The first is reserved for the nails that have the warts around them. The second is used for the rest of the nails that are wart free.

## DRY HANDS
## FEWER WARTS

Just like you should try to keep your shoes as dry as possible, you should also do the same with your hands. This goes a long way in preventing warts!

# Don't Shave!

Well, let me refine that.  Don't shave areas that have warts.  You also want to avoid clipping the areas as well -- at least with the same instruments you use for the rest of your body.

Along the same vein, you want to use a different brush on the areas that have warts than those that are wart-free.

## Avoid Stress

Sounds crazy?  Could the appearance of warts actually be related to stress?  Well, there certainly is a logical progression to that conclusion, according to Dr. Levine.

First, she says she's noticed that many individuals do seem to get more warts when they're under pressure and stress.  Many of these individuals are not following good eating habits at this time either.

Again, we return to the cause.  That virus.  If your immune system is suppressed and under duress because of stress, you're more vulnerable to colds and viruses. It only makes sense that it would include the viruses that cause warts.  Here's one more reason to eat well, exercise more and stay healthy!

If stress is your problem (and it really is for so many of us these days!), then remember to eat foods high in vitamins A, C, and E.

And here's one more tip.  It's not the stress itself that causes the warts, but your reaction to stress!  Learn how to intelligently manage your response to the stressors of your life and you'll not only be helping your warts, but your entire system.  One way?  Try yoga.  These ancient postures help to relax and calm you.

While moles are generally harmless, there are those times when they  may be a symptom of something more serious, like skin cancer. In this case, you don't want to remove them yourself.  You want a trained physician to look at it, remove it and send it off for analysis. How can you tell if the mole you're staring at

right now is "regular" or "irregular"?  That's
what the following chapter is all about.

# Chapter 6:
## When Good Moles Go Bad:
## Is It Skin Cancer?

It's not a pleasant topic. And certainly not one many of us like to think about much. But it does deserve a mention. What am I talking about? Skin cancer. More specifically, I'm honing in on the ability to be able to recognize a suspicious mole.

And, no, this suspicious mole will not be wearing a burglar's mask and carrying a big bag to carry his loot away in. But he may stand out in a crowd of other moles. And by recognizing him and getting him examined by a doctor, you are doing yourself a big favor.

We've all been told that swift action is one of the best ways to increase your odds of treating cancer -- any type of cancer -- most efficiently. But, the only way to do that is to know what you're looking at.

But first, let's explain in just a few paragraphs, what skin cancer is. Then you can better understand why your moles may be your first alarm system to notify you about the development of this disease.

## WHAT IS SKIN CANCER, AFTER ALL?

Skin cancer is technically described as an abnormal growth of skin cells. This growth, in most instances, appears on areas of your body that have had the most exposure to the sun. But, even saying this, know that it may also show up in areas that do not see the light of day much.

The medical community recognizes three types of cancer. They're basal cell carcinoma, squamous cell carcinoma and melanoma.

# Development of
# Skin Cancer

For most people the **first signs** of skin cancer appear on the **face, ears, neck, chest, arms and hands.** For some, the initial symptoms even appear on the scalp or the lips, believe it or not.

But many people are surprised to learn that its first appearance can also be in some of the most unexpected areas. The **palm of your hands. Underneath your fingernails. Even in the spaces between your toes.** For some the first indications even appear in **the genital area!**

Skin cancer can affect anyone of with any skin tone, even those with a darker complexion. So if you've been sunbathing smugly looking at your fair-complexion friends feeling badly because they're wearing hats and long sleeves, you better reconsider your stance.

# Symptoms of
# Basal Cell Carcinoma

Basal cell carcinoma, for the most part, can be found on areas of your body that have been most exposed to the sun. If you develop either of the two changes on your skin listed below, then visit your physician as soon as possible.

- Pearly or waxy bump.
- Flat, flesh- or brown-colored scar-like lesion

## Squamous Cell Carcinoma

This type of skin cancer, squamous cell carcinoma, also appears more often than not, on areas that see the sun the most.  This type of cancer is characterized by the appearance of:

- A firm, red nodule
- A flat, scaly lesion whose surface is crusted over

# Melanoma

Melanoma is the type of skin cancer that can initially develop anywhere on your body -- even those areas well hidden from the sun.  And it's the kind that can take a good mole and turn it bad.

In other words an existing, normal-looking mole can become cancerous.  Surprisingly, your gender has a direct link to the likelihood of where on your body this cancer might strike.

For men, the areas most often affected first, include their trunk, head or neck.  For women, melanoma seems to appear first on the lower legs.

And while it certainly can appear on areas that have been sun exposed, this isn't always the case. Many times, the first signs and symptoms show up in those sun-protected areas.

And here again, don't think just because your skin has a rich, dark complexion that you're immune to melanoma. You may indeed find yourself with it.  You may want to look first on the palms of your hands for this is where it seems to appear first.  For those of you with darker skin, it might also appear under your fingernails or toenails.

So what does it look like you ask?  The following should give you a very good idea of what you're looking for.

- **A large, brown spot with darker speckles on it.**
- **Any mole that changes color, size or bleeds**
- **A small lesion with an irregular border. Portions of this are blue-black, blue, red or even white.**
- **Dark lesions on palms or fingertips, soles of your feet or your toes.**
- **Dark lesions on the mucous membranes of your mouth, nose, anus or vagina.**

# The Types of Skin Cancer

In a very brief explanation, skin cancer is caused when the DNA of health skin cells mutate or reproduce with errors. These mutations then cause the mutated cells to grow at an alarmingly fast rate, which form a mass of what we eventually call cancer cells.

You've noticed that I've identified the various types of cancer by different cells. These cells correspond to the various cells of your epidermis -- or the thin layer of protective covering for the rest of your skin cells.

**Squamous cells**, for example, are those lying just beneath the outer surface of your skin. They perform the duties of being your skin's inner lining.

**Basal cells** are those which produce new skin cells. They're located right beneath the squamous cells.

And the third type of cell is the **melanocytes.** These particular cells produce melanin, the pigment which provides your skin its color. These cells are located in the lower part of the epidermis.

Melanocytes actually produce more melanin when you're out in the sun. They do this in an attempt to protect your deeper layers of skin. And now you can figure out why you tan when you're in the sun, the extra melanin is responsible for the darker color of your skin.

Exactly where your skin cancer initially starts then is the determining factor of its type as well as your treatment options.

## TELL ME AGAIN . . . WHAT CAUSES SKIN CANCER?

We all know that the ultraviolet rays of the sun are among the major contributing factors to the development of skin cancer. I can't help but think that our forefathers knew this instinctively. It's only been recently that we've become such sun worshippers.

Have you ever noticed that farmers have historically worn long sleeves and hats to protect themselves from the sun?
Don't think that because you catch your rays from a tanning bed through a lamp, that you're immune from the development of skin cancer. Because you're not. You're as much at risk as the individual who sits on the beach.

This, however, doesn't explain why areas of your body develop cancer where the sun normally doesn't shine, so to speak. In these cases, it seems there must be some type of toxic substances at work. Or you could have a weakened immune system.

# Who Is More At Risk?

That's a valid question. And in a nutshell, here's a quick list of individual characteristics that might put you at greater risk than your neighbor for developing cancer.

- **Light skin complexion**
- **History of being sunburned**
- **Too much exposure to the sun**
- **Living in a sunny climate**

- **Living at high altitudes**
- **Presence of skin moles**
- **Presence of precancerous skin lesions**
- **Family history of skin cancer**
- **Already having developed skin cancer**
- **Weak immune system**
- **Exposure to toxic substances**
- **Increasing age.**

## PREVENTION IS POSSIBLE

The first and foremost action you need to take is to limit your exposure to ultraviolet radiation. The inner sun goddess or god in you may balk at this suggestion, but you really need to talk him or her into it anyway. It's just that important.

Not only that, but **limit your exposure to the sun during the middle of the day.** This is when the sun is at its strongest. If at all possible stay out of the sun from 10 a.m. to 4 p.m.

Get in the habit of **wearing sunscreen -- every month of the year**. Yep. It sounds strange, but it can help. You and I both know that sunscreens don't filter out all of the radiation. But they are an effective way to filter some of them out.

Play like our forefathers did -- **wear protective cloth-ing.** Cover your skin and wear a broad brimmed hat. This type of hat is even better than either a baseball cap or a visor.

**Just say no -- to tanning beds.** Believe it or not, this only increases your chances of developing skin cancer.

**Watch the medications you take.** Some of them may make your body more sensitive to the sun. Among these are certain antibiotics, drugs taken for cholesterol control as well as high blood pressure. Even nonsteroi-dal anti-inflammatory drugs, like ibuprofen --sold as Ad-vil and Motrin -- may make your skin more sensitive to the sun.

There you have it. A quick overview of what to watch out for in your moles that might indicate a need for medical attention. Not only that, but you get a real feel for the serious nature of skin cancer. Please, if you suspect anything, don't hesitate to visit your physician. It's much better to go early and discover it's nothing than to go when . . . well, you know what I mean!

# Conclusion

Warts. Moles. Skin tags. Just who can keep all these skin growths straight? Ah. You can now! And not only that, you have some of the best tips . . . tricks . . . techniques at your fingertips to get rid of them should you so desire. And as easy as some of these procedures are, why shouldn't you?

But more than that, you have a fuller appreciation of many of the causes of moles and warts and even skin tags. And that's always a good thing.

Remember though, just as the last chapter suggests, that if the mole or the skin growth looks suspicious to you, don't even hesitate to get a medical professional to examine it. Better safe than sorry. There's a reason why that's a cliché. Use it!

I hope this volume has helped you with your skin growth problems in at least some small way. I enjoyed preparing it for you.

You may want to keep a copy of this e-Book handy. Why not set it up as a shortcut on your computer's desktop? This way you always have it available for consultation should you have any lingering questions or if you just want to act like an authority when a friend or neighbor calls for help.

## Here's lookin' at you, kid!

# Resources

## WEB SITES

8 methods for removing skin tags at home,
http://hubpages.com/hub/8-methods-for-removing-skin-tags-at-home, 12 Jul 10.

Basic Home Removal of Skin Tags,
http://skincare.lovetoknow.com/Home_Removal_Skin_Tags, accessed 12 Jul10.

Cantharidin for wart treatment,
http://www.webmd.com/skin-problems-and-treatments/cantharidin-for-wart-treatment, accessed 10 Jul 10.

Common warts,
http://www.mayoclinic.com/health/common-warts/DS00370/DSECTION=prevention, accessed 18 July 10.

Cryotherapy,
http://dermnetnz.org/procedures/cryotherapy.html, accessed, 9 Jul 10.

Dandelion Milk,
http://www.google.com/search?source=ig&hl=en&rlz=1G1GGLQ_ENUS387&=&q=dandelion+milk+wart+removal&aq=f&aqi=&aql=&oq=&gs_rfai=CObMzWuY0TNTGO4LeNfLYnZAKAAAAqgQFT9AhZEc, accessed 7Jul10.

Getting Rid of Warts,
http://www.susangaer.com/studentprojects/warts.htm, accessed 7 Jul 10.

Goodheart, MD, Herbert P, Surgical Pearl: A rapid technique for destroying small skin tags and filiform warts, Dermatology Online Journal, Vol. 9, No. 5, http://dermatology.cdlib.org/95/pearls/tags/goodheart.html, accessed 12 Jul 10

Home Remedies for Skin Tags, http://www.myhomeremedies.com/topic.cgi?topicid=409, accessed 13 Jul 10.

Home remedy for wart removal, http://www.learningherbs.com/home_remedy_for_wart_removal.html, accessed 7Jul 10

How To Make An Herb Poultice, http://www.bellaonline.com/articles/art51100.asp, accessed 13 Jul 10.

How to remove skin tags safely and painlessly, http://www.squidoo.com/how-to-remove-skin-tags-safely-and-painlessly, accessed 13 Jul 10.

How to remove skin tags using tea tree oil, http://hubpages.com/hub/How-to-remove-skin-tags-using-tea-tree-oil, accessed 13 Jul 10.

Intralesional immunotherapy for warts using a combination of skin test antigens: a safe and effective therapy, http://findarticles.com/p/articles/mi_m0PDG/is_3_3/ai_n6056500/, accessed 11 Jul 10.

Mole Removal, http://www.emedicinehealth.com/mole_removal/page5_em.htm, accessed, 17 Jul 10.

Remedies for Skin Tags, Moles, Warts, http://www.fatfreekitchen.com/beauty/warts-remedy.html, accessed 13 Jul 10.

Removal of skin tags, http://www.home-remedies-for-you.com/askquestion/24231/removal-of-skin-tags-what-will-help-remove-skin-ta.html, accessed 13 Jul 10.

Removing Moles with Laser Cosmetic Surgery, http://www.laser-treatment.com/removing-moles.asp, accessed 15 Jul 10.

Skin cancer, http://www.mayoclinic.com/health/skin-cancer/DS00190/DSECTION=prevention, accessed 19 Jul 10.

Skin Tag, http://www.medicinenet.com/skin_tag/page5.htm, accessed 12 Jul 10.

Skin tag removal, http://www.targetwoman.com/articles/skin-tag.html, accessed 12 Jul 10.

So Long Skin Tags, http://www.drweil.com/drw/u/id/QAA269352, accessed 17 Jul 10.

Tag, http://www..com/skin_tag/article.htm#what, accessed 12 Jul 10.

Things That Make Bumps In The Night, http://www.thenewhomemaker.com/wartsmolesskintags, accessed 17 Jul 10.

Warts, *www.nlm.nih.gov/medlineplus/tutorials/**warts**/dm04910 3.pdf*, accessed 9 Jul 10.

Warts, http://familydoctor.org/online/famdocen/home/common/skin/disorders/209.html, accessed 11 Jul 10.

Warts, http://www.vashonorganics.com/WSWrapper.jsp?mypage=treatment_symptoms_warts.htm, accessed 7 Jul 10.

Warts: 26 Ways To Win The War, http://www.mothernature.com/Library/Bookshelf/Books/47/137.cfm, accessed 14 Jul 10

Warts Immunotherapy, http://www.warts.org/warts-immunotherapy.html, accessed 11 Jun 10.

Wart prevention, http://www.forces-of-nature.net/topics/warts/wart_prevention.htm, accessed 18 Jul 10.
Wart removal, http://www.warts.org/wart-remover-salicylic-acid.html, accessed 10 Jul 10.

Warts Removal -- Biological Solution, http://www.abateit.com/warts.htm, accessed 8 Jul 10.

Wart Removal by Electrosurgery and Wart Curettage, http://www.warts.org/wart-removal-electrosurgery-and-wart-curettage.html, accessed 11 Jun 10.

Wart Removal with Duct Tape, http://www.drdaveanddee.com/warts.html, 9 Jul 10.
Wart Removals With Lasers, http://www.laser-treatment.com/wart-removal.asp, accessed 9 Jul 10.

What is a skin mole? http://www.wisegeek.com/what-is-a-skin-mole.htm, accessed 11 Jul 10.

# Books

The Doctor's Book of Home Remedies, Rodale Press, Emmaus, PA